MECHANICAL GIANTS

Roland Berry

Hamish Hamilton . London

First Published in Great Britain 1982 by
Hamish Hamilton Children's Books
Garden House, 57-59 Long Acre, London WC2E 9JZ

British Library Cataloguing in Publication Data
Berry, Roland
Mechanical giants.
1. Machinery—Juvenile literature
I. Title
621.8   TJ147
ISBN 0 241 10765 2

Printed in Denmark

Excavator-loader (by JCB)

Here is a vehicle you must have seen lots of times, digging up a road or working on a building site or farm. It has two buckets, one at each end, which enable it to do all sorts of jobs—digging trenches, bulldozing, loading rubble into dumpers, to name just a few. Here it is digging up a road so that new gas pipes can be fitted.

The engine is 70 horse power (h.p. for short), which means it is about as powerful as a fast sports car. It has twelve gears, nine forward and three reverse. Although it is nearly 8 metres long and over 3 metres tall, it is quite small compared to most of the machines you will see on the following pages.

'Big Geordie' (by Bucyrus Erie)

Big Geordie shifts 100,000 tonnes of soil every day, using a bucket which is big enough to hold two large cars and still have room for someone to walk around in it. Big Geordie is so enormous that if it was put on a football field, it would stretch the full length of the pitch!

Instead of moving around on wheels, Big Geordie walks on two huge feet, each measuring 10 metres by 2.75 metres, at a speed of 0.25 k.p.h. You can probably walk thirty times faster than that. Despite its slow speed, Big Geordie needs a 6,250 h.p. engine—that's a hundred times more powerful than most cars.

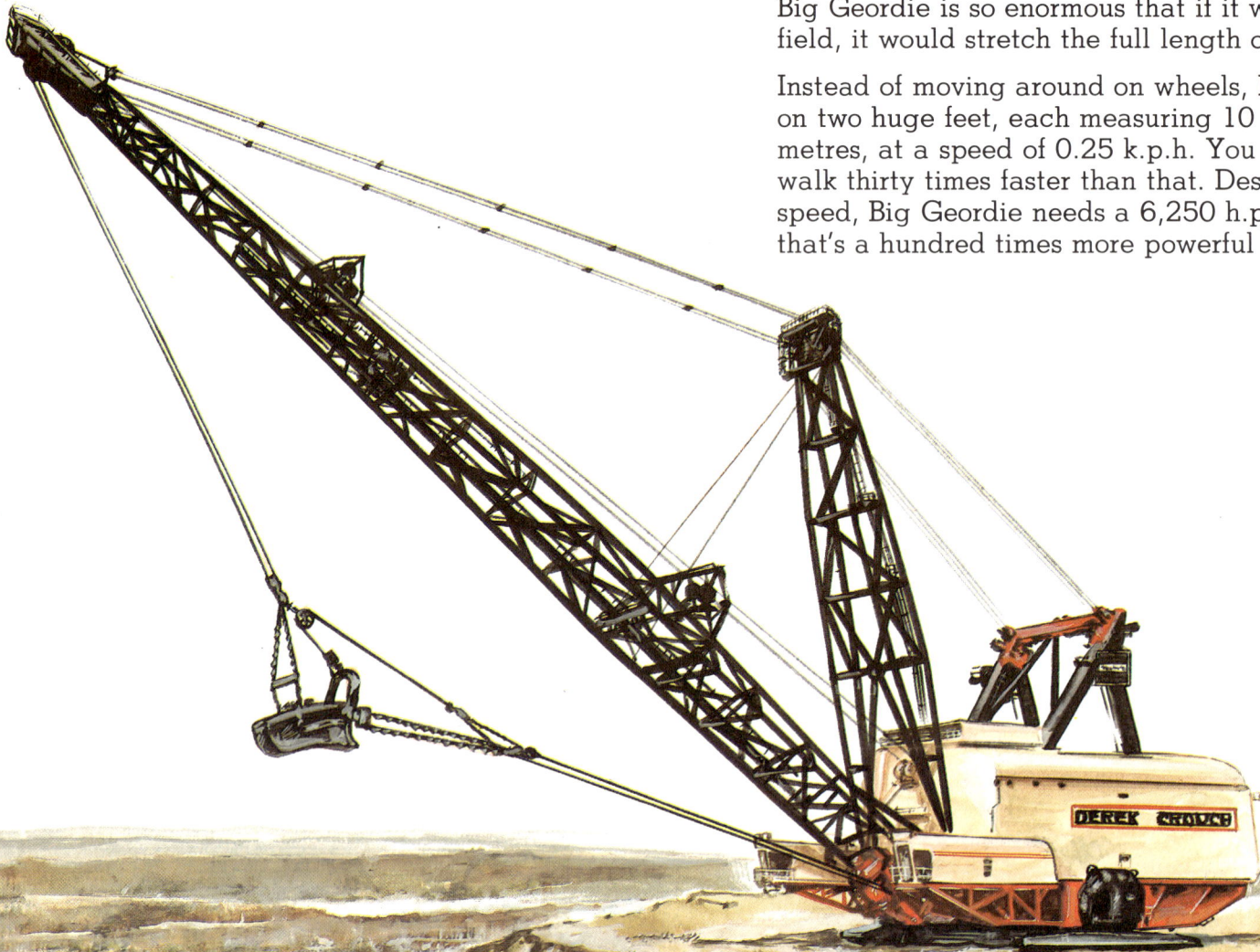

Pipe-mover

The state of Arizona in the USA never has
very much rain. Because it is so dry, the
Colorado River has recently been diverted
from Havasu, on the Californian border,
to central Arizona. Each section of the huge
concrete pipe, designed to carry 1,200 billion
litres of water, is big enough to fit forty cars
inside it. This picture shows the gigantic
vehicle which had to be built to move the pipe.

Diving Bell (by Aqua Logistics)

Diving Bells are lowered from ships and oilrigs to allow divers to descend quickly to depths of over 500 metres. In depths of up to 200 metres, the divers can leave the bell through the hatch at the bottom. If they do not need to leave the bell, they can descend another 300 metres.

The bell is suspended from the mother ship by an 'umbilical', which is a very strong steel cable carrying a hot water hose to keep the divers warm, electrical power for lights and TV cameras, a telephone, and 'Heliox', the gas divers have to breathe at these depths.

Observation submersible (by Comex)

When the divers do not have to leave their underwater vehicle, they can use a submersible like this one. This Comex sub. can move around the sea-bed for up to ten hours at a time without an umbilical.

Both the bell and the sub. help salvage wrecks and maintain oilrigs. This sub. has a long manipulating arm, strong headlights and TV cameras to help the divers in their work.

The 'Jim' atmospheric diving suit

This diving suit is named after the pioneer diver Jim Jarratt, who was the first person to use one like it. It is made of steel several centimetres thick and weighs half a tonne.

Because 'Jim' is so strong, it can be used at a depth of over 600 metres, a depth at which most other diving suits would crumple like paper. It has cleverly engineered oil-filled joints, similar in many ways to the joints in our bodies, which enable the diver to move around on the sea-bed at normal atmospheric pressure, breathing normal air. The diver has much the same freedom as an astronaut walking on the moon.

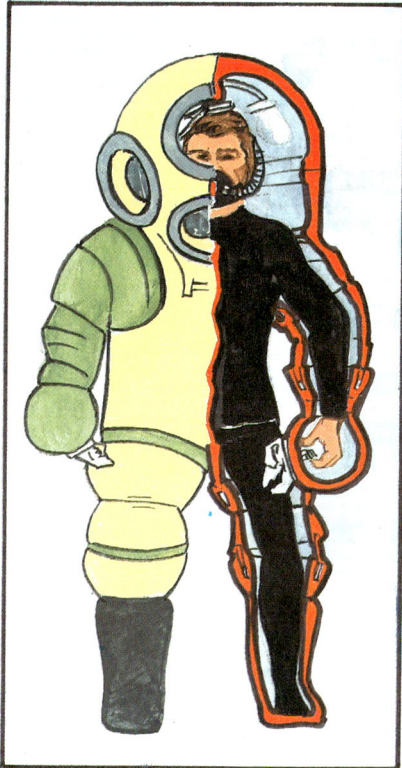

'Brent B' oil production platform

Look at the helicopter which is about to land and you will see how large this production platform, or oilrig, really is— especially when you realise there is even more of it below the waves than above!

The three concrete legs go down to a huge concrete base resting on the sea-bed. This gives the oilrig enough stability to withstand waves up to 25 metres high and winds of up to 200 k.p.h.

The hundred men living on this rig do anything from drilling for oil or repairing equipment, to cooking for the others or painting the superstructure with special anti-corrosion paint to stop the sea water from corroding the metalwork.

'Bendy' tractor (by International)

This International 3588 tractor bends in the middle, or to be more accurate, is articulated. This means that it has a very tight turning circle and can keep all four wheels on the ground, even when the ground is very rough.

With its 180 h.p. turbo-charged diesel engine and twenty-four gears (sixteen forward, eight reverse), it is an extremely versatile vehicle for ploughing, planting and harvesting on larger farms.

It is also very comfortable to drive, and is fitted with a stereo tape player and AM/FM and CB radio.

Combine harvester (by New Holland)

Farmers have only a short time in which to harvest cereal crops such as wheat or barley. Nowadays, they use combine harvesters to help them do the job more quickly. These can cut the corn, thresh it and winnow it and then load the grain into trucks without a break.

Big Baler (by Howard)

The big baler is used to bind up straw or hay into large rectangular bundles, or bales. It can cover nearly 3 hectares in an hour, making bales which measure $2.5 \times 1.5 \times 1.5$ metres (which is about thirty times bigger than normal bales). Each one weighs about 700 kilogrammes.

As the bales are so big, tractors with special grips front and back have to be used to carry and stack the bales. Even then, a tractor can only carry two bales at a time.

Coffee-picker (by Jacto)

Picking coffee is a hot, slow and backbreaking job when it's done by hand. It's a bit like picking blackberries, but worse. This machine can do the same job as a hundred men and women, but much more efficiently.

Trundling down the rows of coffee bushes, the picker beats and shakes the coffee 'cherries' off the bushes with a series of paddles. It then drops the 'cherries' into a container at the back. Hydraulic rams above the wheels move up and down to keep the coffee-picker level, even on very rough ground.

Crop sprayer (by Rockwell)

Large fields of crops are often sprayed with insecticide to protect them against insects such as aphids or locusts, or against diseases such as potato blight. The fastest way to do this is to use an airborne crop sprayer.

This is a Rockwell Thrush 600 Commander, one of the world's largest crop sprayers. It can carry nearly 1,500 litres of liquid insecticide or, with slightly different equipment, 1 tonne of fertilizer granules.

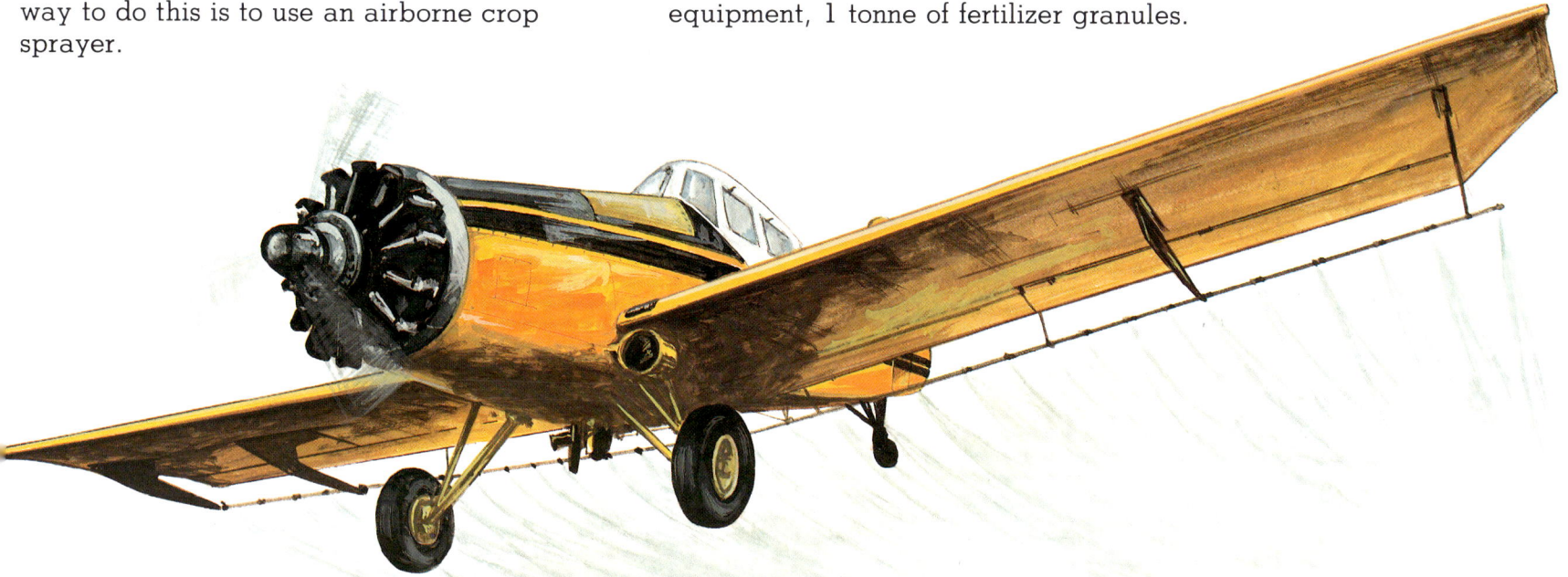

Logger's truck (by Butler Bros of Victoria

250 tonnes of logs can be carried in one load on this Canadian-designed logging truck, one of the biggest in the world. The driver sits in the squashed-looking cab alongside a V-16 Diesel engine.

Tree-chopper (by Caterpillar)

In countries such as Canada, where forestry is big business, special machines have been developed to cut down and transport trees. In the top picture, you can see a tree-chopper. Basically, this is a Caterpillar tractor with a cutting tool at the front. The tool has two 'arms' which are positioned either side of the tree and which then cut down the tree in a matter of minutes. Notice the strongly-built cage over the tractor which will protect the driver should the tree fall the wrong way.

Bulldozer (by Caterpillar)

When hundreds of tonnes of rubble and soil have to be moved in quarries or on construction sites, a larger than average bulldozer is sometimes used.

This Caterpillar D10 bulldozer is one of the biggest. It is so big that if it has to be moved from one site to another it may have to be dismantled and transported on four large articulated lorries. It is 8 metres long and 4.5 metres high and uses a 700 h.p. diesel engine.

Loader (by Caterpillar)

Articulated vehicles such as this are used to load rubble from quarries or construction sites into trucks. This 988B Caterpillar loader weighs 39,200 kilogrammes and has a 375 h.p. diesel engine. It can lift 6 cubic metres of rubble in one bucketful (that's over 15 tonnes). If that seems a lot for one load, its larger brother, the 992C, is twice the size and can lift twice as much.

The dumper truck, which is also in the picture, looks quite large, but it is a midget compared to the Wabco 3200.

Scraper (by Terex)

When a major road is being built, or before a building is erected, vast amounts of soil may have to be moved to make the ground smooth and level. This is where the scraper comes in.

This massive machine, with an extremely powerful engine at each end, scrapes off the top layer of soil into its bucket and carries it away to dump it elsewhere.

Even with the two engines, developing over 700 h.p., it is sometimes necessary to use another scraper (or a large bulldozer) to push it along if the soil is very hard.

Dumper truck (by Wabco)

Surprising as it may seem, the wheels of this gigantic dumper truck (one of the world's largest) are driven in exactly the same way as an electrically-powered milk float, i.e. by electric motors built into the hubs of the two front wheels. As it is rather larger than a milk float, these motors do not run off batteries, but are powered by a diesel locomotive engine, which drives a generator to supply the motors with electric power.

This Wabco 3200 dumper truck can carry more than 200 tonnes of rock or coal. When the back is tipped up, it stands higher than a four-storey building.

Motorway bridge crane (by Sparrow)

Occasionally, machines have to be specially built to
cope with one particular problem.

Special cranes had to be built to lift 160 concrete
beams into place on this motorway bridge. Each beam
weighed 76 tonnes. The two halves of the crane lifted
a beam up to the right height then, moving along
rails, carried the beam to its correct position. When
one span was completed, the crane had to be
dismantled and reassembled on the next span.

Road surfacer (by Barber-Greene)

When new roads are being built or old ones repaired, one of the last stages of the job is to put down the top layer of asphalt and aggregate (gravel or crushed stone). As roadworks almost always cause traffic hold-ups, the quicker this is done, the sooner the traffic will be moving normally again. This Barber-Greene surfacer can put the top layer on a two-lane road at the rate of 16.7 metres per minute. The asphalt/aggregate mix is tipped into the hopper at the front and then spread out across the road, leaving a beautifully smooth surface up to 30 centimetres thick.

Puller-pusher (by Scamell)

Heavy loads are often transported by train. But when something as large as this distillation vessel has to be moved, it has to go by road. You might have seen one of these huge loads crawling slowly along a motorway. It seldom travels faster than 15 k.p.h. and often drops below walking pace when it has to climb hills.

Although this load weighs 350 tonnes, the trailer is capable of carrying more than twice that amount. To spread the weight, there are 224 wheels on the trailer and 10 more on each of three Scamell tractor units, making a total of 254!

Australian road train (by Kenworth)

Very big vehicles are often needed to move large amounts of fuel, food and other commodities. Australia is a country with huge open spaces and very long straight roads, so it is possible to use 'road trains', which have anything up to seven trailers. Because of the length of these road trains and the amount of dust they kick up, the driver may use close-circuit television, instead of mirrors, to keep an eye on the road behind. Most road trains are equipped with CB radio so that the drivers can keep in touch with each other on their long journeys. A round trip can be as much as 8,000 kilometres.

The huge bumper bars, or 'roo bars', at the front fend off kangaroos which often leap out in front of the road trains.

American custom truck (by Kenworth)

In America, many long distance trucks are owned by their drivers. They are often built to the driver's specifications and may cost more than a house to buy.

Some of the drivers spend a lot of time and money making their 'rig' look different, with beautiful pictures, bold stripes and masses of chrome.

As the trucker may be away from home for weeks at a time, the cabs are kitted out like a caravan with bunks, a cooker and even a sink.

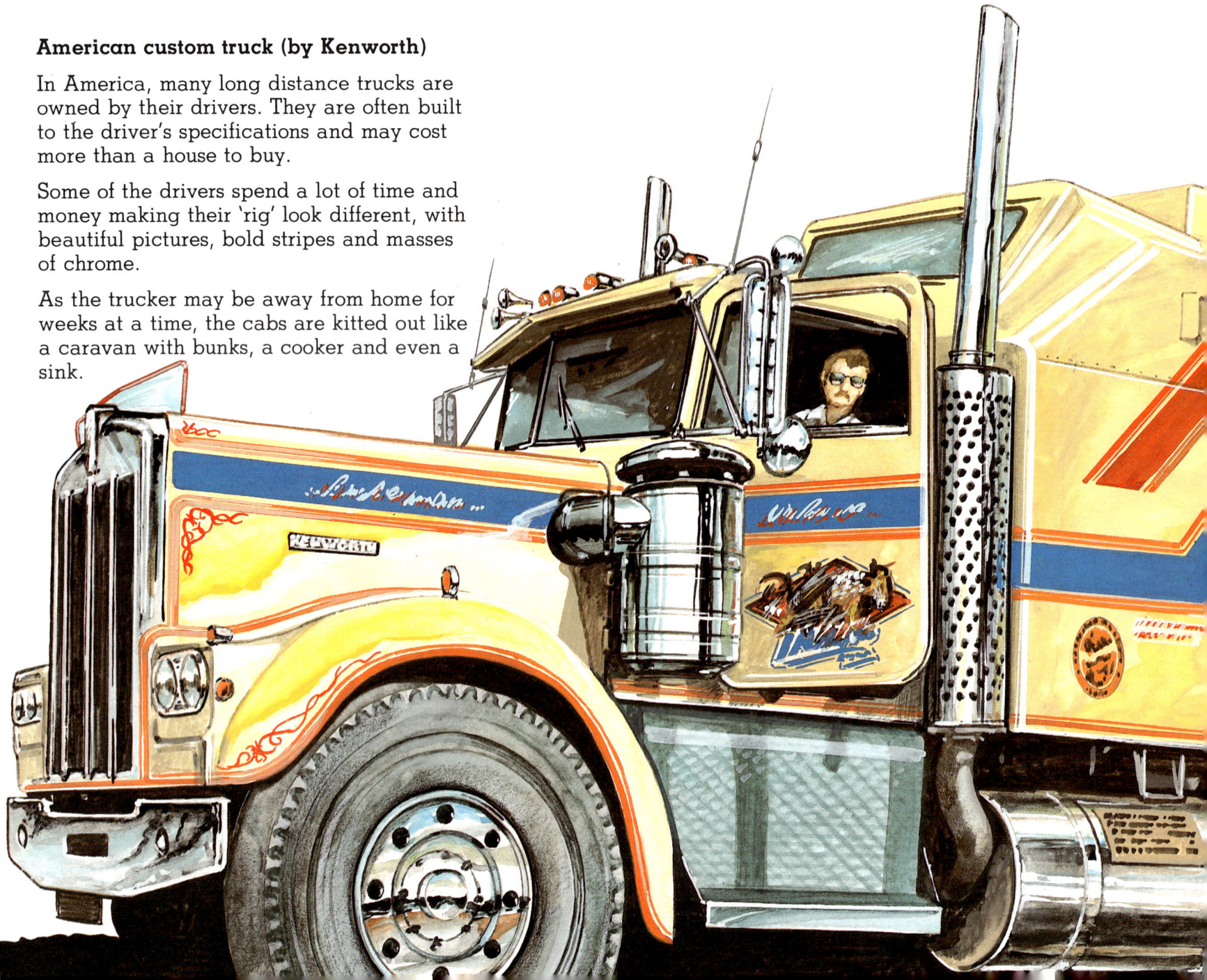

Piggy-packer (by Canadian Pacific)

Moving large loads is always expensive, so if one stage in the handling of the cargo can be cut out considerable savings can be made. The piggy-packer does just this.

Normally, a truck arrives at the docks and its cargo is unloaded onto pallets and lifted by crane into the hold of the ship. The piggy-packer simply lifts the trailer section of the articulated lorry off the ground and puts it on the ship. The tractor unit of the lorry then drives off to get another tractorload of cargo.

Canadian Pacific coal train

It is difficult to believe that this train, which carries coal 1,100 kilometres across Canada, is 2.4 kilometres long and, fully loaded, weighs 1,400 tonnes.

Obviously, an enormous amount of power is required to pull this load; it is supplied by twelve locomotives which are spaced all down the train, altogether developing 36,000 h.p. One driver drives all these locomotives from the front of the train by remote control.

You can imagine how long it would take to unload the coal from all 108 cars, if they were emptied in the usual way. But Canadian Pacific have introduced a huge circular dumper to tip each car upside down to empty it and then turn it upright again, without decoupling it from the rest of the train. The whole operation takes less than three minutes for each car. Even so, you can easily work out that it still takes about five hours to unload the whole train.

Space shuttle crawler-transporter (by Bucyrus Erie)

The space shuttle is the largest space ship ever to have been put into orbit, so it's not surprising that the machines which service it have to be pretty big too.

One of the most impressive pieces of equipment is this 3,000-tonne crawler-transporter, which was originally built for the Apollo space mission. It transports the space shuttle from hangar to launch pad at a speed of 1.6 k.p.h. yet only employs a crew of fifteen.

There are so many astounding facts and figures about the space shuttle that it's difficult to know where to start.

On lift-off it uses 4 tonnes of fuel per second and burns the grass around the launch pad over a distance of 2 kilometres. Its computers, which help with lift-off and landing, perform 325,000 calculations per second (ten times faster than on Apollo missions). The pressure in the engines is 700 times that of a household pressure cooker. And can you imagine flying at 28,000 k.p.h.?!

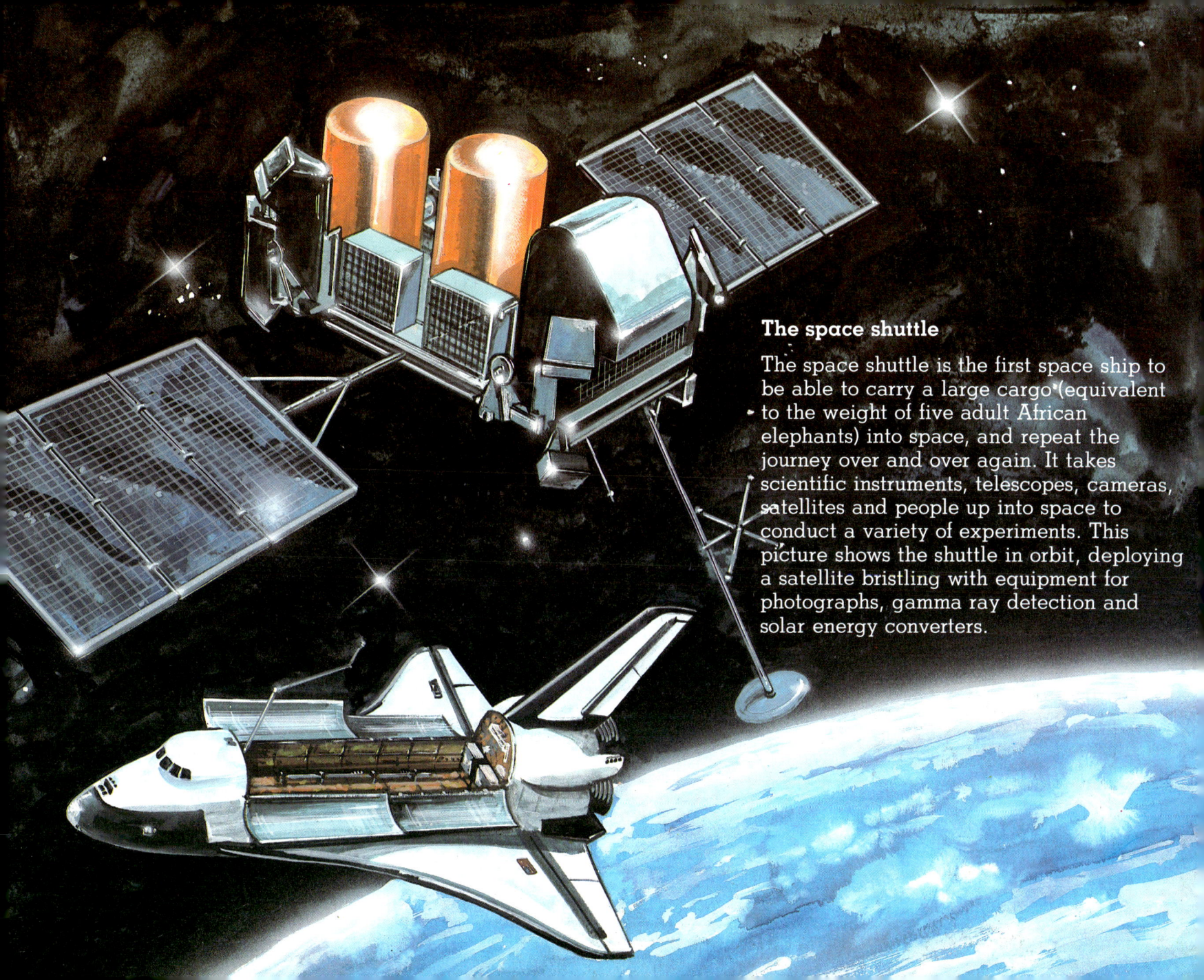

The space shuttle

The space shuttle is the first space ship to be able to carry a large cargo (equivalent to the weight of five adult African elephants) into space, and repeat the journey over and over again. It takes scientific instruments, telescopes, cameras, satellites and people up into space to conduct a variety of experiments. This picture shows the shuttle in orbit, deploying a satellite bristling with equipment for photographs, gamma ray detection and solar energy converters.

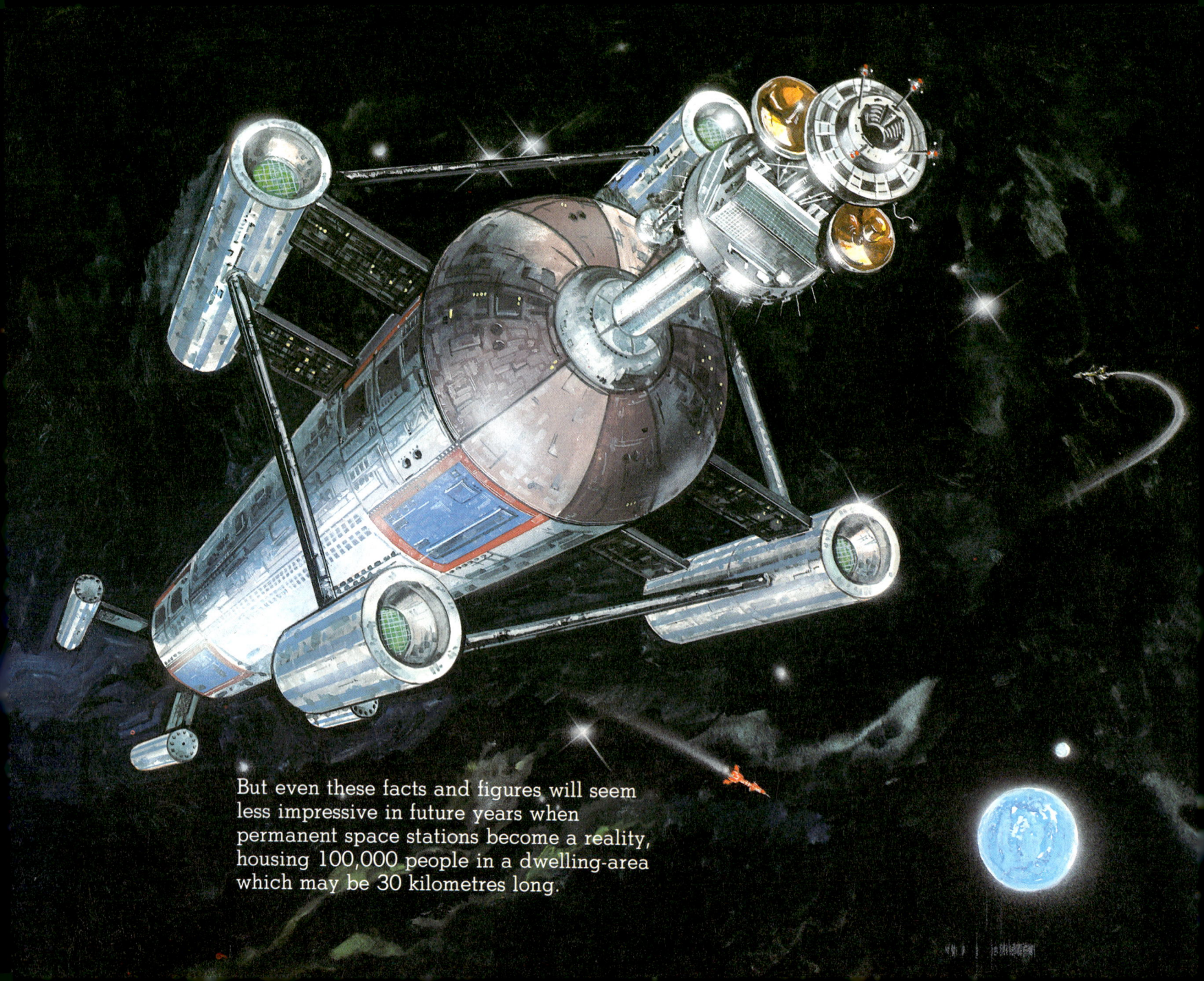

But even these facts and figures will seem less impressive in future years when permanent space stations become a reality, housing 100,000 people in a dwelling-area which may be 30 kilometres long.